Home Geography for Primary Grades

By

C. C. Long

Published by Forgotten Books 2012

PIBN 1000055324

HOME
GEOGRAPHY

FOR PRIMARY GRADES

BY

C. C. LONG, Ph.D.

AUTHOR OF NEW LANGUAGE LESSONS, LESSONS IN ENGLISH, ETC.

AMERICAN BOOK COMPANY

NEW YORK   *CINCINNATI*   *CHICAGO*

81701

TO THE TEACHER.

GEOGRAPHY may be divided into the geography of the home and the geography of the world at large. A knowledge of the home must be obtained by direct observation; of the rest of the world, through the imagination assisted by information. Ideas acquired by direct observation form a basis for imagining those things which are distant and unknown.

The first work, then, in geographical instruction, is to study that small part of the earth's surface lying just at our doors. All around are illustrations of lake and river, upland and lowland, slope and valley. These forms must be actually observed by the pupil, mental pictures obtained, in order that he may be enabled to build up in his mind other mental pictures of similar unseen forms. The hill that he climbs each day may, by an appeal to his imagination, represent to him the lofty Andes or the Alps. From the meadow, or the bit of level land near the door, may be developed a notion of plain and prairie. The little stream that flows past the schoolhouse door, or even one formed by the sudden shower, may speak to him of the Mississippi, the Amazon, or the Rhine. Similarly, the idea of sea or ocean may be deduced from that of pond or lake. Thus, after the pupil has acquired elemen-

3

tary ideas by actual perception, the imagination can use them in constructing, on a larger scale, mental pictures of similar objects outside the bounds of his own experience and observation.

To effect this, the teacher should visit with her class places where the simpler geographical features in miniature may be observed. If the school is in the city, pupils may be taken to the parks for this purpose. If out-of-door study be impossible, they may be induced to recall objects which they have seen on their way to school or on short excursions in the neighborhood. In the case of children who have little opportunity for observing nature, a drawing, a photograph, or a model will be helpful in giving them a proper idea of the matter. It must not be forgotten, however, that actual observation by the pupil is necessary to seeing clearly and intelligently.

Vegetable and animal life are essential features of the geography of the world, and considerable time should be given to the study of those within the observation of the pupils. Information concerning plants may be gained by outdoor study ; also by planting seeds in boxes and having pupils carefully watch their germination and growth.

Pupils should be encouraged to make collections of the minerals and rocks of their region. These should be classified and arranged for use, not for show.

The lessons about rain, snow, dew, etc., should be given at appropriate times. A wet day will suggest a lesson on rain, a snowy day a lesson about snow. No attempt should be made at "science" teaching, so-called. All that should be sought is to get the pupil thoughtfully to observe, and thus to awaken his interest in the world about him.

Lessons should be conversational in form, which is always a most pleasing style for children, as it is the most natural. The work of the teacher is to awaken and stimulate interest, not to impart information. The attention of the child should be directed to what lies around him. He must observe, and think, and express his thoughts. Nor should his observations be confined to the school and school hours. He should be encouraged to obtain his information by his own searching, without guidance, and report the results.

The development of clear mental pictures is stimulated by expression. "Expression is the test of the pupil's knowledge." Hence, the child should be required to reproduce what he has learned. He may do this by modeling, drawing, and oral and written description. These are placed in the order which should be followed in the training of children.

The inclination of nearly every child left to his own mode of development is to make, in some plastic material, what he has seen. Trying to fashion the hills and valleys with which he is familiar excites his interest, and leads to closer observation. This may be followed by the reproduction in molder's sand, or in clay, of the forms seen in pictures or learned from description. Definitions of the various forms, hill, mountain, valley, island, etc., should be developed as they are molded. The memorizing of definitions should seldom be required, and should never be made a test of the pupil's knowledge.

Reproduction by the hand should be followed by drawing, whenever this can be done. Drawing teaches the child how to see well. It often enables him to reveal what could not well be expressed in words. He also becomes ready and rapid in the use of the pencil when he has ideas to put on paper. Only reasonable accuracy should be required. Practice in making fine pictures should not be the end sought, but the development of geographical ideas.

Finally, pupils should be led to give clear and connected statements of what has been learned. For a language lesson, a written description may be prepared, illustrated by a drawing.

CONTENTS.

LESSON   PAGE

I.—Position 9

II.—How the Sun shows Direction, . . . 10

III.—How the Stars show Direction, . . . 14

IV.—How the Compass shows Direction, . . 15

V.—Questions on Direction, ... 17

VI.—What the Winds bring (Poem), . . . 20

VII.—How to tell Distance, ... 21

VIII.—Pictures and Plans, ... 23

IX.—Written Exercise, ... 27

X.—God made them All (Poem), ... 29

XI.—Plains, ... 31

XII.—Hills, Mountains, Valleys, ... 35

XIII.—Rain, Wind, and Snow, ... 40

XIV.—How Water is changed to Vapor, . . 42

XV.—How Vapor is changed to Water, . . . 44

XVI.—Dew, Clouds, and Rain, ... 46

XVII.—The Fairy Artist (Poem), ... 49

XVIII.—How Rivers are made, ... 50

XIX.—More about Rivers, ... 54

XX.—The Brook (Poem), ... 58

XXI.—Work of Flowing Rivers 59

XXII.—Waterdrop's Story, ... 62

XXIII.—The River (Poem), ... 69

XXIV.—A Map, ... 71

XXV.—Forms of Land and Water, ... 73

XXVI.—More about Forms of Land and Water, . 76

7

LESSON	PAGE
XXVII.—A Trip to the Highlands,	80
XXVIII.—Spring (Poem),	86
XXIX.—Useful Vegetables,	87
XXX.—Useful Grains,	88
XXXI.—Fruits,	93
XXXII.—Useful Plants,	95
XXXIII.—Forest Trees,	99
XXXIV.—Flowers,	103
XXXV.—What is Necessary to make Plants grow	105
XXXVI.—Summer Rain (Poem),	107
XXXVII.—The Parts of Animals,	108
XXXVIII.—The Covering of Animals,	110
XXXIX.—Uses of Animals,	111
XL.—The Signs of the Seasons (Poem),	115
XLI.—Things found in the Earth,	116
XLII.—More about Things found in the Earth,	123
XLIII.—How People live, and what they are doing,	127
XLIV.—More about what People are doing,	137
XLV.—A Review Lesson,	141

HOME GEOGRAPHY.

Lesson I.

POSITION.

Lay your hands upon your desk, side by
side.

Which side shall we call the right side?
The left side?

Put your hands on the middle of your desk

on the side farthest from you. That part is
the back of your desk.

Think which is the front of your desk.
Put your hands on the front of your desk.

Who sits on your right hand? On your
left? At the desk in front of you? At the
desk behind you?

Turn round. Who is on your right now?
On your left? Before you? Behind you?

Turn again. Who is now on your right?
On your left? Before you? Behind you?

NOTE.—Lead children to see that the terms *right*, *left*, *front*,
and *back* are of little use in telling the position of places, and
that some fixed standard of direction is necessary.

Lesson II.

HOW THE SUN SHOWS DIRECTION.

If I should ask, "Which is the way to
your home?" who could tell me what I
mean?

You all know which way you must go
to find your home, but if you should wish to
go to a place where you have never been,

you would ask, perhaps, "Which way is it?"

The way to a place is called *direction*. In order to find a place, we must know in what

"THE WAY TO A PLACE IS CALLED DIRECTION."

direction from us it lies, and we have names for directions, such as *north, south, east*, and *west*. We may know these directions by seeing where the sun is.

Did you ever see the sun rise? Point to the place where you saw the sun rise. The direction in which the sun seems to rise is called the *east*.

Did you ever see the sun set? Point to

where you saw the sun set. The direction in which the sun seems to set is called the *west*. The west is just the opposite direction from east.

When do we see the sun rise? Where do we see the sun rise? What is the name of this direction? When do we see the sun set? Where do we see it set? What is the name of this direction? On which side of the schoolroom does the sun rise? On which side does it set? Which is the east side of your desk? Which the west side?

When coming to school this morning, in what direction did you see the sun? If we walk so that the morning sun shines in our faces, in what direction are we going? What direction is behind us?

Now that you know the east, you will be able to find other directions in this way: Stretch out your arms so that your right hand points toward the east, and your left hand toward the west. You are now facing the *north*. The direction behind you is the *south*.

"YOU ARE NOW FACING THE NORTH."

Write the following on your slates:

The sun seems to rise toward the east, and set toward the west. The west is just the opposite direction from the east.

When my right hand is pointing to the east, and my left hand to the west, my face is toward the north and my back is toward the south.

ORAL EXERCISES.

Which is the north side of the schoolroom? Which is the south side? Who sits to the north of you? To the south?

In what direction do the pupils face? On which side of your schoolroom is the teacher's table? Which sides have no windows? Which sides have no doors?

If a room has a fireplace in the middle of the east side, which side of the room faces the fire? Suppose the wind is blowing from the north, in what direction will the smoke go?

In what direction from the schoolhouse is the playground? What is the first street or road north of the school? The first street or road east? South? West?

In what direction is your home from the school? The school from your home? The nearest church from the school? The post office from your home?

Lesson III.

HOW THE STARS SHOW DIRECTION.

You have learned how to tell north, south, east, and west by the sun; but how can we tell these directions at night?

Ask some one to point out to you a group of seven bright stars in the north part of the

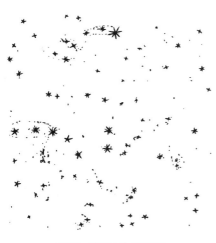

THE GREAT BEAR.

sky. Some people think that this group of stars looks like a wagon and three horses; others say that it looks like a plow.

The proper name of the group containing these seven stars is the Great Bear. The group was given this name because men at first thought it looked like a bear with a long tail.

These seven stars are called the Dipper. It is a part of a larger group called the Great Bear. Find the two bright twinkling stars farthest from its handle. A line drawn through them will point to another star, not quite so bright, called the North Star. That star is always in the north; so by it, on a clear night, you can tell the other directions at once.

Write on your slates:

Sailors out on the sea at night often find direction by looking at the North Star.

Lesson IV.

HOW THE COMPASS SHOWS DIRECTION.

But there are times when it is cloudy, and neither the sun nor the stars can be seen. How can we tell direction then?

Have you ever seen a compass? It is a box in which is a little needle swinging on the top of a pin. When this needle is at rest, one end of it *points to the north.*

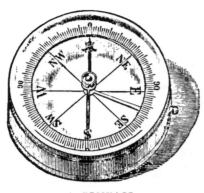

A COMPASS.

As the needle shows where the north is, it is easy to find the south, the east, or the west.

With the compass as a guide, the sailor, in the darkest night, can tell in what direction he is going.

North, south, east, and west are called the *chief points* of the compass.

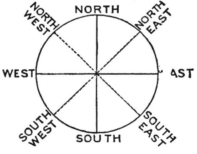

POINTS OF THE COMPASS.

Other directions are northeast, halfway between north and east; northwest, halfway between north and west; southeast, halfway between south and east; and southwest, halfway between south and west.

Write on your slates:

The chief points of the compass are north, south, east, and west.

Other directions are northeast, southeast, southwest, and northwest.

Sailors find their way over the ocean by the help of the compass.

Lesson V.

QUESTIONS ON DIRECTION.

Your teacher will give you time to discover answers to these questions. She could tell you, but it is better to find them out for yourself.

If I go out of doors, how can I find the north? How can I find it on a starlight night? How can I find it on pleasant days? How on rainy days? How does a sailor find the north?

If you were lost and knew your home was north, how would you find it? Do you know how hunters and Indians who live a great deal in the woods find out where the north is? When you are in the woods, notice the amount of moss on the north side of trees as compared to that on the south side.

As winter approaches, many of our birds

will want to go to a warmer country; in what direction will they fly? Point to where ice and snow have their home. What direction is that?

In what direction does your shadow fall at sunrise? At sunset? At noon? When, during the day, is your shadow shortest?

"IN WHAT DIRECTION DOES YOUR SHADOW FALL?"

In what direction does your shadow extend from yourself when it is shortest?

What time of day is noon? How can we tell when it is noon? When is the sun highest in the sky?

"WHAT MAY WE DISCOVER BY WATCHING THE SMOKE?"

What may we discover by watching the direction of the smoke from the chimneys? What does a vane on a steeple tell us? What is a north wind? A south wind? An east wind? A west wind?

What kind of weather may be expected from a north wind? From a south wind? From an east wind? From a west wind?

Lesson VI.

WHAT THE WINDS BRING.

Comes the north wind, snowflakes bringing
   Robes the fields in purest white,
Paints grand houses, trees, and mountains
   On our window-panes at night.

Hills and vales the east wind visits,
   Brings them chilly, driving rain;
Shivering cattle homeward hurry,
   Onward through the darkening lane.

Heat the south wind kindly gives us;
   Reddens apples, gilds the pear,
Gives the grape a richer purple,
   Scatters plenty everywhere.

Flowers sweet the west wind offers,
   Peeping forth from vines and trees;
Brings the butterflies so brilliant,
   And the busy, humming bees.

Each wind brings his own best treasure
  To our land from year to year;
Blessings many without measure
  E'er attend the winds' career.

LILLIAN COX.

"Whichever way the wind doth blow,
  Some heart is glad to have it so;
And blow it east or blow it west,
  The wind that blows, that wind is best."

Write all that you can tell about the wind.

What was the direction of the wind during the last snow-storm? Why is the north wind cold? Why is the south wind warm?

Lesson VII.

HOW TO TELL DISTANCE.

To tell where a place is, we must know its direction. But this is not all; we must also know how far it is from us; that is, its *distance*. To find this out we measure.

You have often heard of an *inch*, a *foot*, and a *yard*. This line is one inch long |. Your ruler is twelve inches long, that is a foot. Three lengths of your ruler make a yard. A yard stick is three feet long.

MEASURING SHORT DISTANCES.

With these measures you can tell how long your slate or your desk is, or how long and wide the schoolroom is.

The inch, foot, and yard are used for measuring short distances. But when we wish to tell the distance between objects far apart, we use another measure called a *mile*. A mile is much longer than a yard.

Think of some object that is a mile from our schoolhouse. How long

MEASURING LONG DISTANCES.

would it take you to walk that distance?

ORAL EXERCISES.

How many inches long is your slate? How long is your desk? How many feet long is your room? How wide is it? What is the distance around the room? How many feet wide is each window? Each door? How many yards wide is the nearest street or road?

About what is the height of the schoolroom? Of the schoolhouse? Of the tallest tree near by? Of the nearest church spire?

About how long is the longest street in the town where you live? Do you know how many blocks or squares make a mile? Name the nearest river or creek. Give its direction from the school. In what direction does the water run? Give the direction and distance of the nearest church. What must you know to go to any place?

NOTE.—Have pupils estimate distances by the eye, then verify by actual measurement. Continue the exercises until the work becomes quite accurate. Correct ideas of distance are necessary in order to understand how large the world is, and how far apart places are on its surface.

Lesson VIII.

PICTURES AND PLANS.

You all know what a picture is. But do you know what a plan is?

A little boy wanted to show his cousin, who lived some miles away, the shape and size of his house, and how the rooms were arranged. How could he do this?

On a large sheet of white paper, he placed
lines of blocks in the form of his house.
Then, with a lead pencil, he drew a line on
the paper along the sides of the blocks. He
next took up the blocks, and there, on the
paper, was a plan of his house.

"THE PICTURE SHOWS THE OBJECTS."

Here is a picture of a schoolroom. We
see desks, the teacher's table, a chair, a clock,
globe, and two maps, in the picture. The
picture shows these objects as they would
appear if we stood at the door behind the
teacher's table and looked in.

This is a plan of the schoolroom, a picture of which is shown on page 24. You see, the plan and picture are quite different.

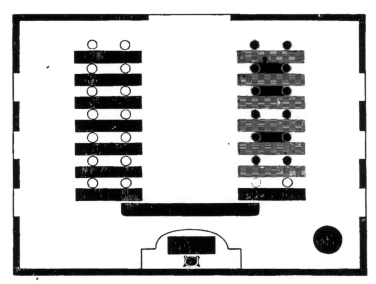

"THE PLAN SHOWS WHERE THE OBJECTS ARE."

The picture shows the objects as we see them before us. The plan shows where the objects are, and their direction from one another.

Now let us see if we can make a plan of the same schoolroom on the blackboard.

The first thing is to measure the sides of the room. We will suppose the two long sides are each forty feet long, and the two short sides each thirty feet long. Now we will draw four straight lines on the board for the four sides. Of course, the lines must be much shorter than the sides themselves, else our plan will be too large.

Make one inch in the plan stand for one foot in the room. So the lines for the long sides will each be forty inches long, and the lines for the short sides thirty inches long.

The next thing is to make spaces in the sides for the door and the windows, and oblongs for the desks. But we must remember that an inch in our plan stands for a foot in the object itself, and therefore we must allow as many inches for the width of doors and windows, and for the length and width of the desks, as there are feet in the objects themselves. Thus, if the door is three feet wide, we must make it three inches wide in our plan.

And lastly, we will draw a circle for the globe, and an oblong and square for the teacher's table and chair, that shall show just where and just how long these objects are.

We have now a *plan* of the schoolroom. Let us put N. to show the north side of the room, S. to show the south side, E. to show the east side, and W. to show the west side. We can now tell the direction of one thing from another in our plan.

Lesson IX.

WRITTEN EXERCISE.

PICTURE OF SCHOOL GROUNDS.

Write the answers to the following questions, in **full** sentences:

What is the name of your school? On

what street or road is it? Which side of
the street? Between what streets? In which
direction does the building face?

How many rooms has the building? In
what part of the building is your room?

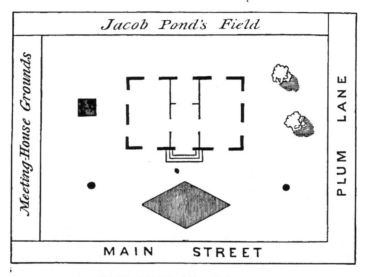

PLAN OF SCHOOL GROUNDS.

How large is it? How many doors and windows? How many seats?

In what direction is the school from your
home? How far is it? How long does it
take you to walk to school?

EXERCISES IN DRAWING PLANS.

Draw a plan of the schoolroom on your slates. It cannot be drawn on your slates as large as it was drawn

on the board. So let one inch stand for ten feet, instead of for one foot; that is, use a *scale* of one inch for every ten feet. Your plan will not be as large as mine, but it will show the position of everything as correctly.

Draw a plan of the top of the teacher's table, showing two books and an inkstand upon it. First, measure the sides. Then decide to what scale you will draw your plan.

Now draw a plan of the schoolhouse and grounds. You must measure not only the house, but the width and length of the yard. The plan must show the size, shape, and place of everything upon the grounds. (While drawing a plan of this kind, it is better to let the pupils face the north. The top of the plan should be the north side of the grounds.)

Draw a plan of your own room at home, showing the table, bed, chairs, and other objects in it.

ORAL EXERCISE.

If the shape of a room is shown on the blackboard, what have we drawn? Is a plan the same as a picture? What is the use of a plan? Mention some things of which plans can be drawn.

NOTE.—It is wrong to teach that the *top* of a map or plan is *always* north; as often as not, the bottom is north, in plans especially.

Lesson X.

GOD MADE THEM ALL.

"THE PURPLE-HEADED MOUNTAIN, THE RIVER RUNNING BY."

All things bright and beautiful,
    All creatures great and small,
All things wise and wonderful,
    The good God made them all.

Each little flower that opens,
    Each little bird that sings,
He made their glowing colors,
    He made their tiny wings.

The purple-headed mountain,
    The river running by,
The morning and the sunset,
    The twinkling stars on high;

The tall trees in the greenwood,
The pleasant summer sun,
The ripe fruits in the garden—
He made them every one.

He gave us eyes to see them,
And lips that we might tell
How great is God Almighty,
Who hath made all things well.

Lesson XI.

PLAINS.

The floor of our schoolroom is level. The playground is almost, if not quite, level. As you look away from the school, is the land nearly level? Did you ever see a broad extent of nearly level land?

Let us imagine that we are out on a piece of nearly level land, many, many times larger than our playground. Such a broad, nearly level stretch of land is called a *plain.*

If this plain were covered with rich green grass and beautiful flowers, we should call it a *prairie.* In the summer it is a vast sea of waving grass. On the prairie we might find herds of wild horses and cattle, which feed upon the rich grass. If it were late in the summer, when the grass is dry and crisp,

"SUCH A BROAD LEVEL STRETCH OF LAND IS CALLED A PLAIN."

it might catch fire, and we might then see a grand sight—a prairie on fire.

We now come to another plain, miles and miles long, miles and miles wide. No rain falls here, and therefore we see no grass, nor flowers, nor cattle, nor horses, nothing but dry, burning sand, rocks, or gravel. We are in a *desert*. But we are so thirsty and tired !

No water to drink, no shade from the burning sun! Suddenly, in the midst of the desert, we come to a beautiful grassy spot. There

"THIS PLEASANT SPOT IN THE DESERT IS CALLED AN OASIS."

is a cluster of date-palm trees, and, better still, a well or a spring of fresh water. This pleasant spot in the desert is called an *oasis*. Here we may quench our thirst, and rest beneath the shade of the trees.

An *oasis* is a fertile spot in a desert. What does *fertile* mean? When do we say land is fertile? When barren? When desert?

Find a picture of a palm tree, and try to draw it.

If we were really in a desert, we might see a company of merchants carrying goods to sell in the countries they visit. Such a company is called a *caravan*. The goods are packed in bundles, which are carried on camels' backs. The camel can live for a long time without drinking, and can carry a heavy load of merchandise a long distance. It is sometimes called the ship of the desert.

Why do travelers use camels to cross the desert? Why do they not use horses? If you can not find answers to these questions in your books at home, ask your teacher about them.

You have seen a small whirlwind in the street. The leaves flew round and round, the dust whirled along in clouds. Trees are sometimes torn from the ground, and houses overturned, by a strong wind.

Now think of a wind-storm in the desert. A loud, rustling noise is heard. Great clouds of fine sand are lifted into the air—clouds which darken the sun! Travelers must at once jump from their camels, cover themselves with their cloaks, and lie flat on the ground.

The poor beasts will close their eyes and nostrils, and kneel with their backs to the wind until the storm has passed over.

Thankful will the travelers be if none of them are buried in the sand.

Lesson XII.

HILLS, MOUNTAINS, VALLEYS.

The land is not always level like a plain. In some places it is high and uneven. We all know what a *hill* is. It is land a little higher than the surrounding country.

Is there a hill near where you live? Let

us walk to the *top*, and stand on its *summit.*
We will start from the *foot* or *base* of the hill.
Now we have climbed its steep, rough *sides*

"WHAT CAN YOU SEE FROM THE TOP OF THE HILL?"

or *slopes.* Was the ascent difficult? Is the
view from the top fine?

What can you see from the top of the hill
—meadow, river, lake, town? What grow
on the hill? What live on the hill?

Which part of a hill is called the base,
or foot? The slope, or sides? The top, or
summit?

Give two names for the lowest part of a
hill. Two for the highest part. Two for the
part between the highest and lowest parts.

Parts of the land very much higher than
the surrounding country are called *mountains.*
Mountains are much higher than hills. Have

you ever seen a steeple one hundred feet high? A mountain is as high as twenty such steeples, one on the top of the other. How high the mountains must be!

Some mountains reach away above the clouds. Their white tops seem to touch the sky. A man on the summit of one saw the clouds beneath his feet, while the sun shone where he stood. When it lightened he saw the flash far below him.

Is it warm or cold at the tops of mountains? With what are many high mountains covered, even in summer?

The land between mountains or hills is called a *valley*. Is there a valley near here? What do you call the ground on either side?

Would you like best to live on the mountains or in the valley? Why?

Are mountains of any use?

Yes, hills and mountains are of very great use. They make the earth more beautiful. Tops of high mountains are so cold that they turn the clouds into drops of water which fall as rain or snow. Then mountains give rise to

rivers which make the valleys beautiful with
grass and flowers.   Mountains do much good
to some countries by keeping off cold winds.

"THINK OF A REAL VALLEY BETWEEN MOUNTAINS."

They also give us coal and iron and other
minerals which we find so useful.

Here is a picture.   What do you call the

very high land on the right and on the left? The long, narrow piece of land between the two mountains?

When you look at this picture you must think of a real valley between mountains.

Bring pictures of hills and mountains to school, if you can find them.

If you had a molding-board and a few quarts of sand, you might represent hills and mountains with valleys between. Think of a real hill while you mold.

Draw on your slate a hill you have seen with a little of the surrounding country.

Write.:

A long, narrow piece of land between hills and mountains is called a valley.

A hill is land a little higher than the country about it.

A mountain is land that rises to a very great height above the country about it.

Lesson XIII.

RAIN, WIND, AND SNOW.

Do you see the dropping rain,
Pitter-patter on the pane?
How it runs along the street!
And it wets our little feet;
But it makes the green grass grow,
And the tiny streamlets flow.

Listen to the wintry blast
Moaning, shrieking, howling past,
Striking with tremendous force
Rocks and forests in its course;
But it blows the windmills strong,
And it sends big ships along.

Watch the pretty snowflakes fall,
Some are large and some are small;
Look, they cover all the ground,
Miles of dazzling white around;
But this covering, I am told,
Keeps the earth from frost and cold.

Ah! and I must work alway,
Life's not meant to spend in play;
Every moment's fleeting fast,
And our day will soon be past;
If our work is truly done,
It will last though ages run.

Of what use is rain? Of what use is snow? Of what use is wind?

Lesson XIV.

HOW WATER IS CHANGED TO VAPOR.

What happens when a kettle of water is put on a hot stove?

The water gets hot and boils away.

Where does it go? Is it destroyed?

The water is changed, but it is not destroyed. Coal burns, but we do not get rid of it altogether. It is changed into gas and smoke and ashes.

What is the water changed into?

It is changed to vapor. If we let the kettle remain on the fire long enough, the water it contains will all pass away as vapor.

Where does the vapor go? The water, though turned into vapor, must be somewhere.

It is floating about in the air of the room, though we cannot see it. The air holds the vapor, just as a sponge holds water.

Heat expands or swells air. Warm air, therefore, can contain more vapor than cold air. On a warm day there may be many times as much moisture in the air as on a cold day.

Moisten your slate with a damp sponge. Observe the disappearance of the moisture.

Dip your hand in water, and wave it in the air. The water on your hand disappears. Where has it gone?

When wet clothes are hung on the line, they soon become dry. What becomes of the water in the clothes?

If we set a plate of water out in the sunshine, what happens? Is the water lost?

The streets and roads were wet and muddy, now they are dry. What has become of the water? Has it all sunk into the ground?

Sometimes we see leaves and grass sparkle with water-drops, early in the morning. When the sun shines out and warms the air, what happens?

Why does vapor rise into the air?

Why does smoke go up? Because it is lighter than air. As vapor is lighter than air, what do you think ought to happen to it?

Lesson XV.

HOW VAPOR IS CHANGED TO WATER.

Heat, as you have learned, changes water into vapor. You must also know that cold turns vapor back into water again.

"THINK OF THE KETTLE WITH THE BOILING WATER."

Now let us think of the kettle with the boiling water. You will notice a little space, quite close to the spout, where nothing can be seen. Is there no vapor there?

Yes, there is vapor there, but it cannot be seen; it is invisible. A little way from the spout we see something white, like smoke. This is only the vapor that has been chilled by the cool air and changed back again into water. The water is in the form of very

fine particles, and may be called water-
dust.

Hold a cold plate over boiling water. Observe how the water-dust gathers into drops
that roll down the plate.

You have seen the inside of windows in
cold weather covered with moisture. Where
does it come from? Why did it form there?
Why does it sometimes run down on the
cold pane?

The vapor in our breath turns into water
on frosty mornings. Explain this.

Carry a pitcher of ice-water into a room,
and notice what takes place. A thin mist at
once gathers on the outside of the pitcher.
What takes place among the little drops of
mist? What becomes of these larger drops?

Where does the water which collects on the
outside of the pitcher come from? Does it
come through the pitcher from the inside?
Would the same thing have taken place if
some other cold object had been used instead
of a cold pitcher?

Write out what you have learned about vapor.

Lesson XVI.

DEW, CLOUDS, AND RAIN.

The sun is all the time heating the water on the land and in the sea, and changing it into vapor, which rises in the air. We cannot see the vapor, but it is in the air around us.

If the vapor in the air is suddenly cooled, a strange thing happens. Some of it quickly changes back into water. You have often seen, in_the early morning, little_drops of water hanging like pearls upon the blades of grass.

Now, where do these drops come from? They come from the air. The vapor in the air floats against the cold grass and leaves, and is cooled and changed into tiny drops of water. We call this *dew*.

Of what use is dew?

If the night is quite cold, the dew will freeze. It is then called *frost*. You have seen the frosty window pane with the beautiful pictures upon it.

Make a picture of the window as you remember it, covered with the pretty things made by the frost.

"WHEN VAPOR RISES HIGH IN THE COOL AIR."

When vapor rises high in the cool air it is turned into very small drops of water or minute crystals of ice, and we can see it floating about in the air. It is then called a *cloud*. Almost any clear day you may see clouds form and then seem to melt away.

You have seen on a blue sky, light, fleecy feather-clouds. They are very high up, and it is very cold where they are. You have also noticed the clouds at sunset with their beautiful colors. As the sun sank lower and lower, how did they change, in shape and color?

When clouds are low down, near the earth, we call them *fogs* or *mist*.

If clouds are cooled, the little particles of water gather into large drops and fall as *rain*. If the drops should freeze in falling, we would call them *hail*.

What shape are the raindrops? Of what use is the rain?

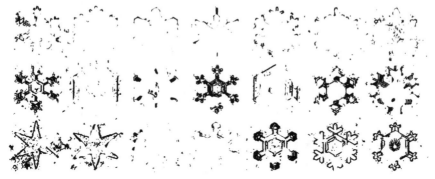

"HAVE YOU EVER SEEN SNOWFLAKES THROUGH A MICROSCOPE?"

Sometimes, when it is very cold, the moisture in the air freezes before it forms into drops, and falls in the beautiful flakes we call *snow*. Have you ever seen snowflakes through a microscope?

Snow keeps the roots of plants warm. Many plants would die in winter if it were not for the snow. What other uses has snow?

Observe the clouds, fog, rain, snow, dew, frost, and tell what you have noticed.

Write what you have *seen* or *noticed* about vapor, clouds, rain, etc.

Lesson XVII.

THE FAIRY ARTIST.

Oh, there is a little artist
    Who paints in the cold night hours
Pictures for little children
    Of wondrous trees and flowers!

Pictures of snow-white mountains
    Touching the snow-white sky;
Pictures of distant oceans
    Where pretty ships sail by.

Pictures of rushing rivers
    By fairy bridges spanned;
Bits of beautiful landscape
    Copied from elfin land.

4

The moon is the lamp he paints by;
His canvas the window pane;
His brush is a frozen snowflake;
Jack Frost the artist's name.

Lesson XVIII.

HOW RIVERS ARE MADE.

Have you ever seen a brook or creek? A river? Is there a brook or river near here? Who can tell where it begins? where the water comes from that fills it? where it goes? Let us try to understand this.

As vapor rises into high, cool air, or is carried with the air in winds up the sides of mountains, it turns into water again, and comes falling down as rain.

Now think where the rain that falls on mountains must go. Some of the water runs off on the surface, down the mountain slope. Some sinks into the ground, and runs along in little streams below the surface. It will appear again, bubbling out of the mountain side as a *spring*. The spring is the beginning of a river.

Did you ever see a spring? Where was
it? Was it shaded by trees? Where did
the water
come from?
Did you
drink

"DID YOU EVER SEE A SPRING?"

from it? Was the water pure and cold? Where
did the water go after leaving the spring?

From the spring flows a tiny, thread-like stream, so small that we can easily step across it. This little stream is called a *rill*.

Other rills meet this, and form a larger stream, which is called a *brook* or *creek*.

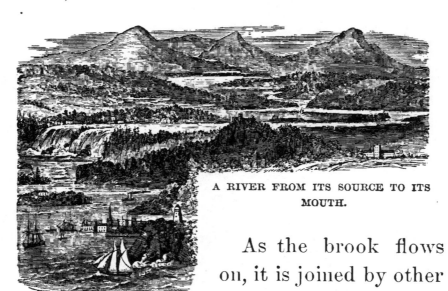

A RIVER FROM ITS SOURCE TO ITS MOUTH.

As the brook flows on, it is joined by other streams, until, little by little, it becomes a wide and deep *river* on which large boats may float. At last, it finds its way into the ocean.

Where a river begins is its *source*. The place where it flows into another body of water is called its *mouth*. The land over which it flows is its *bed*.

A river has two banks. As we go toward

its mouth, the right bank is on our right hand, and the left bank is on our left.

Do you live near a river? Where does the water come from? In what direction does it flow? Why does it flow in such direction? Does it wind about much? Does it flow into the ocean, or into another river?

Is the water fresh or salt? What grow on its banks? Near which bank do you live?

Make a picture of a spring, and a brook flowing from it. Draw the tall grass and plants that grow near it.

Write the names of all the rivers you have seen.

Write the following:

Water flowing out of the ground is called a spring.

From springs flow small streams called rills, brooks, or creeks.

A large stream of water flowing through the land is called a river.

A small stream of water flowing into a larger one is called a tributary.

The source of a river is where it begins. The place where it empties into another body of water is its mouth.

Every river has two banks—a right-hand bank and a left-hand bank.

Lesson XIX.

MORE ABOUT RIVERS.

Let us have another chat about the river. We may fancy that we are following it in its course to the sea. We shall then learn for ourselves many things we do not know about rivers.

We will begin our journey at its source. Here it is a little rill, formed by water that trickles from a spring, or by the melting of snow.

As it flows on, it is joined by many other little streams, until it grows to be much larger.

There is a large word used for a stream that feeds another stream. Do you know what it is? The word is *tributary*. Tributaries are often called *branches*.

Before we leave this part of the river, I wish you to learn another hard word.

You have seen the water run off the roof of a *shed*. The ridge, or highest part of the roof, divides the rain that falls on it, so that part of the rain flows down the one side, and part of it flows down the other side.

Now, hills, like the roof of a house, send off

"HILLS SEND OFF STREAMS ON BOTH SIDES."

streams on both sides. When it rains, or the snow melts, some of the water goes down on one side, some on the other. And that is why the hills which divide or part the waters of streams are called a *water-parting* or *water-shed*.

Let us now go further down the stream.

Here we see it rushing rapidly down a

steep slope.   Its waters foam and dash between the great rocks that lie in the stream. Such places in the river are called *rapids*. Can you tell why they are so called?

"SUCH PLACES ARE CALLED RAPIDS."

The stream flows on. It has now reached a high ledge of rock. Over this it leaps, making a great foam and noise.

When the water of a river falls over high rocks, it is called a waterfall or *cataract*.

You may have seen the Falls of Niagara, the greatest waterfall in the world.

"YOU MAY HAVE SEEN THE FALLS OF NIAGARA."

The course of our river is now through a lower country. The valley in which it flows spreads out, and the stream grows wider and wider. The water moves slower and slower.

Why is the river swift in some places, and in others slow?

At length it flows through an almost level country. It is here widest and deepest. Its course is more winding.

Do you know why it is crooked and winding?

Because on the steep hillside the water runs

very rapidly, and is not easily turned aside
Where the ground is nearly level, it runs
slowly, and is easily turned from its course.

Lesson XX.

THE BROOK.

From a fountain
In a mountain,
Drops of water ran
Trickling through the grasses;
So our brook began.

Slow it started;
Soon it darted,
Cool and clear and free,
Rippling over pebbles,
Hurrying to the sea.

Children straying
Came a-playing
On its pretty banks;
Glad, our little brooklet
Sparkled up its thanks.

Blossoms floating,
Mimic boating,
Fishes darting past,
Swift, and strong, and happy,
Widening very fast.

Bubbling, singing,
Rushing, ringing,
Flecked with shade and sun,
Soon our pretty brooklet
To the sea has run.

Lesson XXI.

WORK OF FLOWING RIVERS.

Would you like to know more about
brooks and rivers—about the work they do?

Notice what happens when it rains. Little
tiny streams are formed, which chase each
other down the slopes. See how they cut
away the loose soil and carry it off. Notice

how muddy this loose soil makes the water.
What becomes of this loose soil, or mud?

Fill a jar with water. Put in a handful of
mud from the nearest stream. Shake the jar,
and the water is muddy. Let it stand awhile.
What do you notice? The water is clear,
and the soil has settled to the bottom.

Follow the streams to the valley where
they unite to form a river. When does the

"THESE FERTILE MEADOWS WERE FORMED OUT OF THE LOAM."

load of mud it carries settle? Here, where
the water scarcely moves, we find some of

the soil spread out over the ground near the river banks.

You have seen a river overflow its banks. When the water went down, it left a layer of rich mud, which made the soil very fertile.

Have you never seen the low ground on the banks of rivers covered with rich grass and clover?

Well, these fertile meadows were formed out of the loam that has been washed down the streams from the far-off hills and mountains.

Look at the jar again. Which settled first, the coarse material or fine loam? What kind of a deposit will be made in the upper course of a river? What kind toward the mouth?

High up in the valley, when the river is low, we see *pebbles* in its bed; lower down, the pebbles are worn into *gravel;* and as we get still farther down, we find the gravel ground into *sand.*

Examine the stones found along the shore of a brook or river. Some are quite smooth and round. They were not always so, but had sharp edges.   Do you know what made them round?

When there are heavy rains, the rushing water sweeps large stones down the mountain side and into the valley.   As they are carried down the stream, the stones, by rubbing against each other, are smoothed and rounded and ground into pebbles.   The pebbles themselves are ground at last into gravel and fine sand.

This is what the streams are doing everywhere—plowing deep furrows in the sides of the mountains, grinding the pebbles and sand into fine soil, and carrying it into the valleys below.

Lesson XXII.

WATERDROP'S STORY.

Patter, patter, fall the raindrops on the brown leaves in the woods. Mr. Squirrel's bright eyes sparkle as he peeps out of his queer little home, a hole in the tree; his store of nuts has been carefully hidden away.

Splash comes a drop on a leaf just opposite him. Such a friendly little drop it is, for soon it tells this little woodland dweller of all its travels.

Let us listen, for we may hear too:

"My home," began the Waterdrop, "is in the wide blue sea, where I live with many, many other drops.

"One day as we rode up and down on the big waves, the sun shone down on us, and we grew warmer. Each little drop felt, 'Oh, if I could only get away from the other drops, how much cooler I should be!' Then each tiny drop separated from the others, and grew so small you could not see it.

"We, of course, grew lighter, lighter than the air. Up, up we rose into the bright blue sky. When we got pretty high, where the air was cool, we came closer together

again and formed a great fleecy white *cloud*, that cast its shadow over everything. Then a friendly wind carried us along, and soon we left the sea behind. Far below, we could see green fields and waving woods."

"You must have been very happy," said the little squirrel.

"Yes; it was a merry life we led, as we floated hither and thither, playing with the sunbeams," replied the Waterdrop.

"But we came at last to a purple mountain, and a chill wind began to blow. How we shivered with the cold! Then we huddled close together to get warm. We were now heavy again—so heavy that we could not stay up in the air.

"Then,

'I'm going down to cheer a flower,'
  Cried a little drop of rain;
'I hear it sigh. It droops its head
  As if in weary pain.'

'And I will go !' 'And I !' 'And I !'
  Cried all the raindrops near.
So down we went in merry haste
  The whole wide field to cheer.

"The drooping flowers lifted their bright faces to thank the little drops for the cool drink. Even the great tall trees nodded their heads in welcome."

" The grass on the hillside and in the valley must have been grateful, too, for your coming," said the squirrel. "It always looks so fresh and green after a shower. But, tell me, what became of *you ?*"

"I fell where the ground was brown and bare, stopped for a moment, then went down, down into the ground, where all was dark. I met other drops trying to get out, and we went on together, turning first this way, then that way, till we burst into the sunshine again.

"We rested for a moment in a tiny pool of clear water; then I ran with the rest down the mountain side, slipping over smooth pebbles, and tumbling over sharp rocks, until I found myself in a deep, swift stream, where plants and trees grew on either bank.

"SUDDENLY WE FELL OVER THE ROCKS."

"As I was hurried along, I heard a great roaring noise made by the river falling over a high ledge of rocks, as a cataract or waterfall. Suddenly we fell over the rocks so steep and high that we went leaping and dashing in all directions. We rose in the air in a fine gray mist, then sank back again into the foam-covered stream.

"Soon we were in a broad, quiet river, flowing past the grassy hills and green pastures. Then we came to a big mill-wheel, upon which we jumped, and by our weight made it turn over and over, and thus move the machinery in the mill. Here we were tossed

"THEN WE CAME TO A BIG MILL-
WHEEL."

in the air, whirled around, and at last flung back into the river, where we sailed slowly and quietly as before.

"By and by, we saw large boats floating on the water. We passed towns and cities with busy streets and many people; and as our river widened, and we heard the big sea waves dashing against the shore, we knew our brothers and sisters were singing a welcome home.

"WE PASSED TOWNS AND CITIES."

"And now farewell, little squirrel. My story is done, and I must hasten to my home in the sea. Perhaps we shall meet again some day. I may float down to you, a white-winged snowflake, or patter down as I came this time, a tiny raindrop."

Write the following:

The water rises from the sea in vapor.

The vapor is turned into clouds, which fall in rain or snow.

The rain forms rivers, which flow back again into the sea.

Thus the water is always going round and round in its long and curious journey—up to the clouds in vapor, down in rain, back in streams to the place it started from.

Lesson XXIII.

THE RIVER.

" Oh, tell me, pretty river,
　　Whence do thy waters flow?
And whither art thou roaming,
　　So smoothly and so slow? "

" My birthplace was the mountain,
　　My nurse the April showers;
My cradle was a fountain,
　　O'er-curtained by wild flowers.

" One morn I ran away,
　　A madcap, noisy rill;
And many a prank that day
　　I played adown the hill!

" And then 'mid meadowy banks,
　　I flirted with the flowers,
That stooped with glowing lips
　　To woo me to their bowers.

" But these bright scenes are o'er,
　　And darkly flows my wave;
I hear the ocean's roar—
　　And there must be my grave! "

Where have you seen a river like the one spoken of
in the poem?　Are rivers born?　What is meant by
" My nurse the April showers "?　" I flirted with the
flowers "?　Explain the last stanza.

Lesson XXIV.

A MAP.

A drawing made to show a room, or a house, or the school-yard, or even a village, is called a plan.

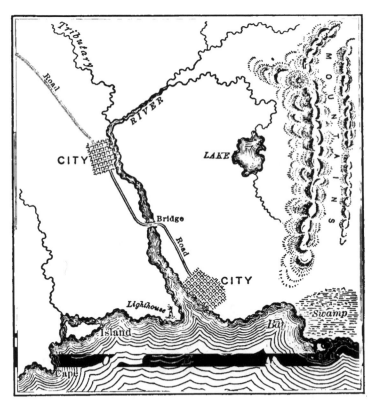

Drawings which represent land and water are called maps. You may learn from maps where the countries, and mountains, and rivers, and cities are that you have seen. It also shows how far places are from one another.

Here is a map showing mountains and rivers. The many short lines facing each other represent mountains. To show the very high part of the mountains, the lines are drawn close to each other, making that part of the map look dark. The line winding about, like the stream itself, represents a river. The line, as you see, is made thicker and thicker toward its mouth. From this you may know that the river itself becomes broader and broader as it flows toward the sea.

But you must not think that the crooked line on the map is a river, or the lines which face each other are mountains. If you do, you will learn very little of geography. When you look at these lines, you must *think* of the real things which they stand for—the lofty mountains, with their covering of forests, and with long, narrow valleys between them; the winding, gently flowing river, bearing boats upon its waters.

Lesson XXV.

FORMS OF LAND AND WATER.

You all know what a pond is. Is there a pond near where you live? Did you ever fish in it? Did you ever walk round it?

When a stream, on its way to the ocean, flows into a basin or hollow in the land, the water spreads out and fills it. A hollow in the land filled with water is called a *lake*, or, if it be quite small, a *pond*.

What is a lake made of? What is round it? Suppose some one who never saw a lake were to ask you what a lake is, what would you say?

What do we find in lakes? Would you not like to sail on a lake?

In the hollows among mountains are great numbers of beautiful lakes. In their clear waters may be seen the mountains, the forests, and the sky, as in a looking-glass. At night the moon and stars may be seen below you as plainly as above.

Here is a picture of a pretty lake in a valley.

You see a river flowing from the hills beyond.   Into what is it flowing?   The river

A PRETTY LAKE IN A VALLEY.

that lets the water *into* the lake is called an *inlet*.

You see another river that lets the water *out* of the lake.   This river we call the *outlet* of the lake.

Make a lake on your molding-board, or in the sand near your home.   Represent its inlet and outlet.

Out in the lake is a little piece of land round

which the waters play. We could not go to this land without crossing the water; the water is on all sides of it. Such a little piece of land is called an *island.*

Did you ever read the story of Robinson Crusoe? You will remember that he went up a hill in search of water. When he got to the top of the hill, he saw that he was on an island. How did he know?

Have you ever seen an island? What island was it? Could you sail round it? What was on every side of it? What grew on it? What is an island?

If there is a brook or lake near your home, how can you make an island?

Opposite is a picture of a river and a lake. Make a map of the same river and lake on your slate. Notice how the coast or shore of the lake bends in and out.

Write the following:

A lake is water surrounded by land.

The land near the water of a lake is called its shore.

An island is a little piece of land surrounded by water.

Lesson XXVI.

MORE ABOUT FORMS OF LAND AND WATER.

PICTURE OF A PENINSULA.

An island, as we have learned, is a piece of land with water all round it. Now, sometimes we see a piece of land that has water *nearly* all round it. This form of land is called a *peninsula*. The word peninsula means *almost an island*.

In the picture we see a narrow strip of land which extends far out into the water. You will notice that the

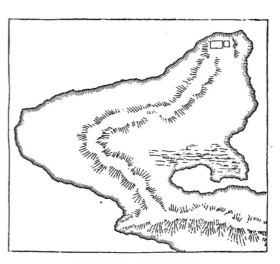

MAP OF A PENINSULA.

land has water all round, except at one place.

What is the name for land that has water on all sides but one? What is a peninsula?

How would you change this peninsula to an island? What is the difference between a peninsula and an island?

The narrow neck which joins the peninsula to other land—just as the neck joins the head to the body—is called an *isthmus*, which means *neck*.

Here is another picture which I wish you to look at. You see where the shore bends like

PICTURE OF A BAY.

a bow, and the water runs a little way into the land.

Can you think of anything else that is bent like this? Yes—a bay-window.

Now, when I tell you that bay means the same as bow, you can almost guess the name for this bend in the land. It is called a *bay*. You will easily remember that little word.

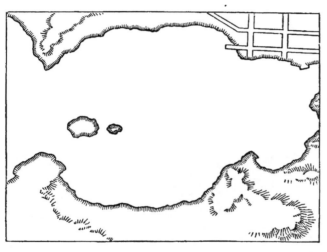

MAP OF A BAY.

A wide opening or bend in the land, into which the water flows, is usually called a bay.

Some-times, when the opening in the bend is long and narrow, it is called a *gulf*.

On the next page is shown a narrow strip of water joining two larger bodies of water. The name given to this narrow passage is *strait*, a word meaning *narrow*.

As an isthmus connects two bodies of land,

so a strait con-
nects two bodies
of water.

After a rain
make little lakes,
rivers, bays, etc.
Perhaps you may
find some already
made.

PICTURE OF A STRAIT.

S e e whether
you can find in the magazines and books at
home pictures of gulfs, bays, peninsulas, etc.

Write the following:

A peninsula is land almost surrounded by water.

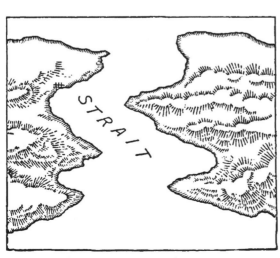

MAP OF A STRAIT.

An isthmus is a
neck of land joining
two larger bodies of
land.

A gulf or bay is a
portion of some large
body of water ex-
tending into the land.

A strait is a nar-
row passage of water
that joins two larger
bodies of water.

Lesson XXVII.

A TRIP TO THE HIGHLANDS.

Uncle Tom had been telling Fred and me about many strange places he had seen. Last of all, he told us about some high mountains he had climbed. We wanted to climb one very much. So father said he would go with us up a high hill not far from the city.

Mother did not need to call us in the morning, for we woke up very early. The sky in the east was bright, and we knew that soon we should see the sun. We wanted to start at once, but mother said it would be better to have breakfast first.

We put on thick shoes that the stones would not easily cut. Father gave each of us a stout stick to help us climb. Fred had a knapsack, in which mother put some bread, cold meat, crackers, and a cup to drink from. In one corner we put some towels.

We were soon outside the city, walking along the road. We passed a village, and

went through fields and woods. By and
by we could see the land before us rising
higher and higher. We saw no longer such
beautiful farms and gardens as we had passed.

"AS WE WENT UP THE SLOPE."

In a little while we reached the foot of the
hill and began to ascend. As we went up the
slope, we came to steep, rugged places that
were hard to climb, where we needed our

6

sticks. The trees were smaller, and there were many bushes. There were large rocks, too, in the sides of the hill. At the foot, the weather was quite warm, but it grew cooler and cooler the higher we went.

"WE COULD SEE THE CITY WITH ITS LITTLE STRAIGHT STREETS."

"On the summit at last!" cried Fred, as we reached the top.

From where we stood, we could see the city with its little straight streets, that look so wide when we walk on them. We could see the house-tops, too, and the church steeples. Then father showed us the village we passed, and the woods we went through. But the trees looked like bushes, and some men at the foot of the hill looked no larger than the baby.

Down the mountain, a little way, we found a spring where the water was clear and cool. Here we sat down on a rock, and ate the lunch we had brought. While we rested, we

"IN THE VALLEY LAY A LARGE SHEET OF STILL WATER."

watched the little rill that flowed from the spring—

"Blue in the shadow,
Silver in the sun."

Down the hill, it was larger, and we saw other rills flowing into it. When it reached the valley, it was much larger; and farther down, father said, boats could sail on it.

Before us, in the valley, lay a large sheet of still water.

"Oh, how pretty!" exclaimed I.

"Yes, that is a lake," said father. "How beautiful it looks as the sunlight plays on its smooth surface! It reflects in its clear water the sky and the trees around it."

"I can see an island in the lake," said Fred. "Vessels are sailing all round it."

"Are all islands small, like that?" he asked.

"Oh, no!" said father. "Some islands are hundreds of miles round, and have many people living upon them."

Fred then pointed to a piece of land extending out into the water, and asked whether that, too, was an island.

"No," replied father, "that is a peninsula. It is land almost surrounded by water. And the narrow neck which joins the peninsula to the mainland is called an isthmus.

"You see the bend in the land, with the water from the lake running in?" asked father.

"Yes," said Fred.

"That is called a bay. Around every ocean, which is a much larger body of water, there are many such bays.

"The narrow strip of water, which a boat is just entering, is called a strait. The strait separates the island from the mainland."

Stretching far away to the east was flat, level land, which father called a plain. Scattered here and there were many farmhouses and quiet villages. Little bright, sparkling streams wound their way like silver threads through the green grass of the meadows. It was a lovely scene indeed!

The sun was already low in the west as we made ready to return. As it set—

A wonderful glory of color,
    A splendor of shifting light—
Orange and scarlet and purple
    Flamed in the sky so bright.

Lesson XXVIII.

SPRING.

Drops of rain and bits of sunshine
  Falling here and gleaming there,
Tiny blades of grass appearing,
  Tell of springtime bright and fair.

Budding leaves are gently swaying,
  Merry glad notes sweetly ring;
Robins, bluebirds, gayly singing,
  Tell of happy, pleasant spring.

Violets, in blue and purple,
  By the twinkling water clear;
Fair spring beauties, frail and dainty,
  Tell the story, spring is here.

Cherry, peach, and apple blossoms
  Scattering fragrance far and wide;
Buttercups and pure white snowdrops
  Tell of gracious, sweet springtide.

LILLIAN COX.

Lesson XXIX.

USEFUL VEGETABLES.

In the heart of a seed buried deep, so deep,
A dear little plant lay fast asleep.
"Wake!" said the sunshine, "and creep
    to the light."
"Wake!" said the voice of the rain-
    drops bright.

A SPROUT.

The little plant
heard, and it
rose to see
What the won
   derful out
side world
might be.

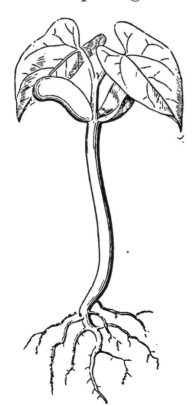

A VINE BEGINNING TO GROW.

What vegetables grow
in your neighborhood?

Of which do we use the
roots as food? Of which
the leaves? Of which the
seeds? Of which the
stems or stalks?

Which is the most useful garden vegetable? There is no common garden vegetable so highly thought of as the potato. How are potatoes planted?

Answer the questions in writing so as to make a little composition about vegetables.

Lesson XXX.

USEFUL GRAINS.

Wheat and corn are called grain because

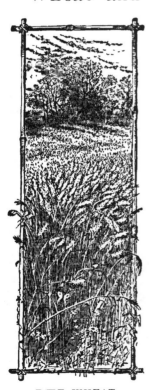

RIPE WHEAT.

they are small, hard seeds. What other kinds of grain can you name?

Which of these grains is used the most? Which makes the choicest flour?

Some kinds of wheat are sown in the spring. These are called spring wheat.

Winter wheat is sown in the fall. A few days of sun and rain, and the plants spring up like grass, remaining green through the winter.

What color does the wheat turn as it ripens? When it is ripe what is done with it? For what is the flour of wheat used?

HARVESTING WHEAT IN THE WEST.

What is sometimes done with the stalks, or straw?

Indian corn is one of the most useful of plants. Do you know why it is called Indian corn? It is because the Indians first raised it.

When is corn planted? How is the land prepared for planting? What is done to the corn while the plants are small? When does it ripen? How tall does it grow?

SEVERAL KINDS OF GRAIN.

What is the stem of the corn called? What are the flowers on the stalk of corn called? On what do the grains of corn grow?

What use is made of the green stalks and leaves? What use is made of the ripe grain? For what are corn-husks largely used?

Sweet corn, if boiled when green, is an excellent vegetable. It is preserved by canning.

A large cornfield, with its tall, straight stalks, covered with green shining leaves and crowned by flowers, is a very pleasant sight.

"ANOTHER GRAIN WHICH WE FIND ON ALMOST EVERY TABLE."

Corn is sometimes called the national emblem. What does emblem mean?

What use is made of oats, barley, rye, and buckwheat? Some of these grains are useful in two or three ways.

There is another grain which we find on

almost every table. It is rice. The rice plant, when growing, resembles wheat; but, unlike wheat, it needs a great deal of moisture. So the rice-grower sows it in fields which he can flood or drain at will.

Do you know what people live on rice without any meat at all? Ask your teacher to tell you how rice is raised in China and Japan.

You ought to find something to tell your teacher and classmates about the grains.

Perhaps you would enjoy drawing some of the grains you have seen.

Choose one of the grains, and write what you have learned about it from conversation and observation.

We plow the fields, and scatter
  The good seed on the land,
But it is fed and watered
  By God's almighty hand.
He sends the snow in winter,
  The warmth to swell the grain,
The breezes and the sunshine,
  And soft refreshing rain.

Lesson XXXI.

FRUITS.

Name some trees upon which grow things to eat. What do we call such trees?

What fruit trees have you seen? What do

"THE ORANGE TREES ARE LOADED WITH GOLDEN FRUIT."

we call the place where many fruit trees grow?

Did you ever pick berries? What makes **it** hard to pick blackberries?

Name fruits that grow about here. Which grow on trees? Which on bushes? Which on vines?

Mention the different uses of these fruits.

The orange is one of the most delicious and wholesome of fruits. It grows only in the warmer parts of our country. In winter as well as in summer, the orange trees are loaded with golden fruit and fragrant blossom. The blossoms are white, and are very beautiful.

Name other fruits that grow in warm parts of the country.

People who live in cold countries need such food as will make them warm. What kinds of food are best in cold countries? What people live mainly on fish and the flesh of animals? Do any fruit trees grow in very cold countries?

What kinds of food are best in hot countries? The people cannot eat fatty food, for that would heat the body. Do we find in such countries grain, vegetables and cooling fruits for the people to live upon?

Write answers to some of the questions asked in the lesson, so as to make a composition about fruits.

Lesson XXXII.

USEFUL PLANTS.

What plant supplies us with much of our clothing? Name articles of clothing made of cotton.

Did you ever see a field of cotton? In the summer the young plant is covered with pretty, pale-yellow flowers. In the autumn you see the pod or boll which contains the cotton.

As the pod ripens, it bursts open. The cotton-field is now a pretty sight —the bright green leaves,

"YOU SEE THE POD OR BOLL."

yellow blossoms, and snowy cotton all mingled

together. Form a picture in your mind of a field of cotton in bloom.

The cotton is now picked. The first thing is to separate it from its seed. This is done by a machine called a cotton-gin.

"FLAX IS A SMALL PLANT."

Now it is ready to be pressed in great bales and sent to market. It will, at last, go to the cotton mills and be spun into thread, then woven into muslin, calico, etc.

Are the seeds of any use? They contain a great deal of oil, which is pressed out by machinery. What is the name of this oil? What use is made of it?

There is another plant from which clothing is made.

Do you know what plant linen is made from? Linen comes from the flax plant.

Flax is a small plant which grows two or three feet high, bearing on the top a bunch of pretty blue flowers. A field of flax in bloom is a very pretty sight.

The flax does not grow in a pod like cotton. The stalk of the plant is covered with a bark, or skin, containing fibers. These fibers are spun into thread, which is woven into a cloth called linen.

The seeds are used for making an oil called *linseed oil*. For what is linseed oil used?

Do you think people who live in hot countries need the same kind of clothing as those who live in cold countries?

A PLANT THAT YIELDS NO FOOD."

What kind of clothing should you think was needed in cold countries? Would such clothes be comfortable in hot countries?

There is a plant that yields no food, drink, or clothing, yet it is used in nearly every country in the world. Can you tell its name?

7

"SUGAR-CANE IS A TALL PLANT."

Every one has seen it growing. It is tobacco.

Do you think the tobacco plant is as useful as the cotton and flax plants?

Everybody eats sugar. Did you ever see a table set for supper without a sugar bowl? .

The sugar in common use in this country is made chiefly from sugar-cane. The sugar-cane is a tall plant which looks much like Indian corn when growing. It is called the sugar-cane because it is filled with the sweet juice that is made into the sugar.

When the stalks are cut they are taken to

a sugar mill. Here they pass between great rollers which press out the juice. The liquid is then boiled until it turns to sugar.

Much sugar is made from the sap of the sugar-maple tree. In the early spring the sap begins to rise. A hole is bored in the tree and a tube inserted, through which the sap passes to a bucket or other vessel placed to receive it. The sap is boiled in large kettles and becomes syrup. More boiling turns the syrup into sugar.

Write what you have learned of *cotton* and *linen*.

Lesson XXXIII.

FOREST TREES.

In your walks what things please you the most? Is it not the trees? Trees are very useful to us, and we ought to be very grateful for them.

Name some trees along the streets and in the parks. Are they useful to us, especially on a hot day? Why? Then what kind of trees do we call them? (Shade.) Which of these are the first to put on their green dresses in the spring? Which are the brightest in autumn?

Name some trees that grow in the woods.

A SHADY STREET.

Name a tree whose wood is dark. A tree whose wood is light. A tree whose wood is hard. A tree whose wood is soft.

Name some trees that are valued for the color and hardness, or the beautiful grain, of their wood.

What kind of wood are the desks made of?
The teacher's table?

What kinds of wood are used in making
chairs? tables? pianos? windows? floors?

If we wish to make a carriage, omnibus,
cart, or wagon, which wood should we use?
Why?

From which trees do we get lumber for
building?

Can you name a wood which is very hard
and tough, and is used in building ships?

What do we call many trees together, like
these?

What is Arbor Day? Why need we plant
trees?

What do we plant when we plant the tree?
We plant the houses for you and me.
We plant the rafters, the shingles, the floors.
We plant the studding, the laths, the doors,
The beams and siding, all parts that be—
We plant the house when we plant the tree.

What do we plant when we plant the tree?
A thousand things that we daily see.

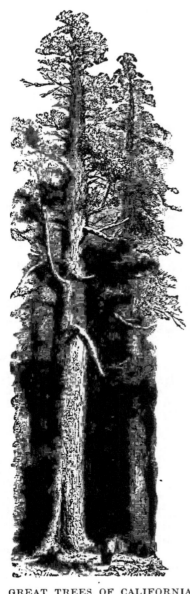

GREAT TREES OF CALIFORNIA.

We plant the spire that
out-towers the crag,
We plant the staff for our
country's flag,
We plant the shade, from
the hot sun free—
We plant all these when
we plant the tree.

There are trees much
larger than any we find
growing here. I am sure
you must have heard of
the great trees of Cali-
fornia. Some of them
are one hundred feet
around, and nearly four
hundred feet high,—
twice as high as a very
tall steeple. In one of
these trees, if hollowed
out, a large family might
live.

In your rambles in the woods, notice and
examine the trees which you see. Learn to

know the trees so that you can call them by their proper names.

Draw and paint some of the objects noticed; as grains, vegetables, trees, etc. You will enjoy this very much, and it will help you to see these things better.

Lesson XXXIV.

FLOWERS.

A flower is a weak and tiny thing; but there are many flowers, and by helping together they cover the earth with beauty and fill the air with sweetness. They seem to have been made to give us pleasure.

It will be easy and useful to learn something about the flowers that grow where you live. How many flowers can you mention by name? Which do you know at sight? Where would you go to find them?

Would you find them all growing in the same place? Which can live only in wet places? Which thrive best where there is but little moisture?

If we take a walk in the fields in the early spring, which flowers shall we be likely to see? Which later? What color are they? Which are fragrant? Which most beautiful? Which would you like for your flower vase? Which would you like to plant and care for in a box of earth or a garden-bed?

Can you find and name the parts of a plant—root, stem, leaves, bud, flower? Learn the uses of each part.

Here are some pretty verses on "Spring and the Flowers." Perhaps you will commit them to memory.

In the snowing and the blowing,
    In the cruel sleet,
Little flowers begin their growing
    Far beneath our feet.

Softly taps the Spring and cheerly:
    "Darlings, are you there?"
Till they answer, "We are nearly,
    Nearly ready, dear.

"Where is Winter with his snowing?
    Tell us, Spring," they say.
Then she answers, "He is going,
    Going on his way.

"Poor old Winter does not love you,
    But his time is past;
Soon my birds shall sing above you—
    Set you free at last."

Lesson XXXV.

WHAT IS NECESSARY TO MAKE PLANTS GROW.

Plants do not grow in winter. Can you
tell why? Plants do not grow in hot places
called deserts. Can you think of any reason
for this?

What two things are necessary to make plants grow? At what time of the year can they get these?

If a country has a great deal of heat and rain, what can we be sure of about its trees and grass and flowers?

There are places that have rain enough, but very little heat. How do you suppose the trees grow there?

You may get information about plants and things by seeing for yourself, by asking others, and by reading books.

"IF A COUNTRY HAS A GREAT DEAL OF HEAT AND RAIN."

Write the names:

Of some grains that we use for food.

Of some vegetables that we use for food.

Of two plants from which clothing is made.

Of woods used in making furniture.

Of woods used in building our houses.

Lesson XXXVI.

SUMMER RAIN.

Oh, gentle, gentle summer rain!
  Let not the silver lily pine,
The drooping lily pine in vain,
  To feel that dewy touch of thine,
To drink thy freshness once again,
Oh, gentle, gentle summer rain!

In heat the landscape quivering lies,
  The cattle pant beneath the tree;
Through parching air and purple skies
  The earth looks up, in vain, for thee;
For thee—for thee it looks in vain,
Oh, gentle, gentle summer rain!

Come thou, and brim the meadow streams,
  And soften all the hills with mist,

Oh, falling dew! From burning dreams
  By thee shall herb and flower be kissed;
And earth shall bless thee yet again,
Oh, gentle, gentle summer rain!

Lesson XXXVII.

THE PARTS OF ANIMALS.

Animals which live in or near people's houses and are tame are domestic animals; others are wild.

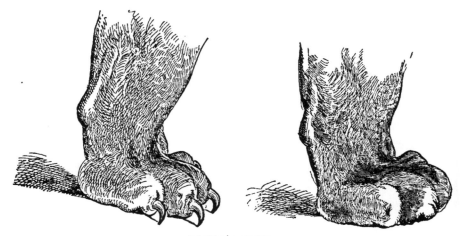

TIGER'S FEET.

What animals are found where you live? Which are domestic? Which wild?

Which of these animals feed upon grass? grains? fruits? flesh?

You know what sharp claws a cat can put out when it pleases. What use does the cat make of its claws?

How does a cat's paw differ from a dog's? How does a dog seize its prey?

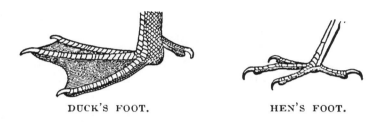

DUCK'S FOOT.   HEN'S FOOT.

Compare a duck's bill with an owl's. What use does the duck make of its broad flat bill? The owl, of its sharp hooked bill?

How do the bills of the hen and the duck differ? Would a bill fitted for pecking be as useful to the duck as its own bill?

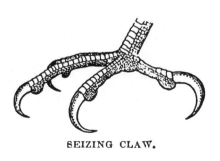

SEIZING CLAW.

Can you draw a picture of a duck's foot and a hen's foot? For what does each use its feet?

Would broad web-feet be as useful to the hen as slender toes?

What kind of feet has the sheep? For what are its feet only used? Could a sheep use feet like those of a cat or a hen?

You see each animal has parts well fitted for the life it leads.

Lesson XXXVIII.

THE COVERING OF ANIMALS.

Name two animals covered with fur.

Two covered with hair. Two covered with feathers.

What do we call the covering of a sheep? Of a pig? Of what use is hair to animals?

What covering has an oyster? A lobster? A turtle? Of what use is it to them?

The duck's feathers are covered with an oily coating, which keeps them from getting wet. Are the feathers of the hen so covered? Why?

Suppose a squirrel's covering were like that of a turtle's, what would result?

What would result if a bird had scales instead of feathers? You see that each animal has that covering which suits its habits best.

Lesson XXXIX.

USES OF ANIMALS.

What domestic animals are used for food?
What wild animals are used for food?

"WHAT DOMESTIC ANIMALS ARE USED FOR FOOD?"

From what animals do we get beef? pork?
mutton? veal?

What birds and fowls are used for food? What fishes?

From what animal do we get wool? How is wool taken from the sheep? What articles of dress are made of wool?

"FROM WHAT ANIMAL DO WE GET WOOL?"

Name the animals whose skins are used to make leather. How is leather prepared? Did you ever see a tannery? What do they do there?

From what animals do we get furs?

What is silk?

Silk is made by little worms called silk-worms. When the worm is fully grown, it spins round itself a small ball of silk, called a

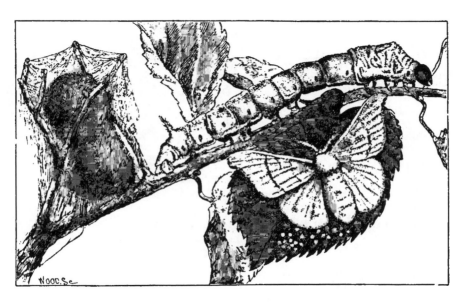

THE SILKWORM AND MOTH.

cocoon. If this cocoon were left to itself, the worm would change to a moth, and the moth would eat its way out of this little house. But this, of course, would cut the little threads and spoil the silk. As soon, there-fore, as the cocoon is made, it is put into hot water to kill the worm. In this way the silk is saved.

Almost every part of the cow is made use of. For what is the flesh used ? What use is made of the hoofs ? horns ? hair ? What is done with the skin ? What other uses has the cow ?

What animal shows the most affection for his master ?

Mention some kinds of dogs.

You may have seen a dog called the St. Bernard. He is large, with long curly hair. In the Alps mountains, where traveling is dangerous, the St. Bernard dogs have saved many lives. Who use their dogs, as we use horses, to draw their sledges ?

Which is the most useful animal to man ?

Draw and paint some of the animals spoken of in the lesson.

Write the names :

Of animals useful for food.
Of animals which work for man.
Of animals useful to us for clothing.
Of birds and fowls used for food.

Lesson XL.

THE SIGNS OF THE SEASONS.

What does it mean when the bluebird comes
    And builds its nest, singing sweet and clear?
When violets peep among blades of grass?—
    These are the signs that spring is here.

What does it mean when berries are ripe?
    When butterflies flit, and honeybees hum?
When cattle stand under the shady trees?—
    These are the signs that summer has come.

What does it mean when the crickets chirp,
    And away to the south the robins steer?
When apples are falling, and leaves grow brown?—
    These are the signs that autumn is here.

What does it mean when days are short?
    When leaves are gone, and brooks are dumb?
When fields are white with drifted snow?—
    These are the signs that winter has come.

Lesson XLI.

THINGS FOUND IN THE EARTH.

The earth contains many things that are of great value to us. These we must find and dig out.

The coal we burn in our grates to warm us; iron, from which so many useful things are made; gold, silver, tin, lead, and copper, —all come out of the earth.

But these are not all the valuable things hidden away in the earth.

From salt wells we get a great deal of the salt used on our tables. From oil wells is obtained the oil we use in our lamps to give us light. Diamonds which sparkle so beautifully, and the stone we use in building, are also taken from the earth.

Coal, iron, gold, silver, lead, tin, copper, mercury, and salt are called minerals.

The opening dug in the earth from which minerals, except stone, are taken, is called a *mine*.

One of the most useful of minerals is *coal*. Did you ever stop to think how much hard work coal does for us? It grinds our wheat, it weaves our cloth, it carries us by sea and by land over the earth. Hardly

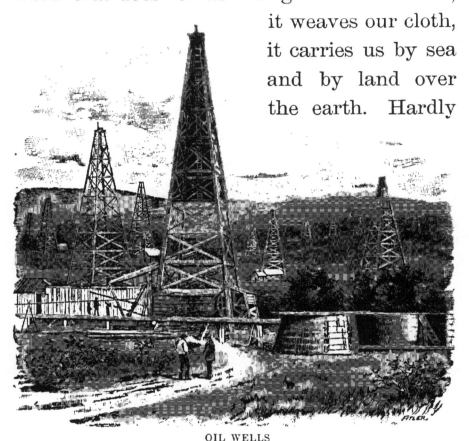

OIL WELLS

any labor can be done without coal.

You have noticed that some coal burns with a great deal of flame and smoke. That is called soft or bituminous coal. That hard, clean-looking coal, which burns with little

blaze, yet gives out such great heat, is anthra-
cite coal.

Coal has many uses. Mention all you can
think of. From which kind is gas obtained,
hard or soft coal?

What is coal? Some day you will be able
to understand how coal was made, and how
it got deep down in the earth.

What article used with food is found in
mines? Does all *salt* come out of the mines?
How is the salt made that is not found in
mines?

There are salt mines where men, women,
and children live all their lives, and never see
sun or sky. Many great rooms and galleries,
with tall pillars to hold up the roof, are cut
out of the salt. When lighted up with
torches, they glitter as if studded with precious stones. It is like a fairy palace.

Some minerals are called metals. Iron,
gold, silver, copper, tin, and mercury are
metals.

Iron is the most useful of all metals. Did
you ever think what we should do without

this hard, strong metal? The following lines tell some of the uses of iron:

Iron vessels cross the ocean,
Iron engines give them mo-
    tion;
Iron pipe our gas delivers,
Iron bridges span our rivers,
Iron horses draw our loads,
Iron rails compose our roads;
Iron houses, iron walls,
Iron cannon, iron balls,
Iron lightning rods on spires,
Iron telegraphic wires,
Iron hammers, nails, and
    screws,
Iron everything we use.

IRON MINE.

Steel is iron made very hard. Knives, axes, hatchets, and other tools are made of steel. Many little things are made of steel. Mention some of them.

Which is the most valuable of all metals? Is all the gold made into money? Is money made of pure gold? Why? Name articles of ornament made of gold. Articles of use.   Are gold watches, chains, and rings usually made of pure gold? Why? What do you call the man who makes these articles?

Silver is the whitest and most lustrous of

CASTING IRON FROM THE ORE.

all the metals.   What does "lustrous" mean? Is iron lustrous? Are silver articles usually made of pure silver? Why?

Silver and gold are found among the mountains in the west. Sometimes they are dug out of the ground. Sometimes they are found in rocks, and the rocks must be broken up before they can be taken out.

Sometimes men wash down the hills with streams of water in order to get at the silver or gold among the rocks.

Gold and silver are called the *precious*

"SOMETIMES MEN WASH DOWN THE HILLS."

metals because they do not rust, and on account of their scarcity.

Tin is white and bright, but too soft to make articles which shall be light and strong. Therefore, thin plates of iron are dipped into melted tin. The tin adheres to the iron and makes it bright like tin itself.

A thin sheet of iron, covered with tin,

is called tin-plate. It is of this that our tin cups, pans, and kitchen utensils are made. A tin cup is really made of iron.

Lead is a very heavy metal. It is so soft that it can be cut with a knife. It is used in making shot, and water pipes.

Do you know how shot is made? Did you ever see a shot-tower? Small shot is made by dropping melted lead through a sieve in rapid motion, from the top of a high tower. The drops fall into a vessel of water below. They are next polished and made black, and then are ready for sale.

You think, I suppose, that the lead pencil with which you write is made of lead. It is not made of lead, but of graphite, which is a kind of coal.

Copper is softer than iron, but harder than lead. It will not rust. Cooking vessels are often made of copper.

Zinc is another valuable metal, and is almost the color of tin. Brass is made by mixing copper and zinc together.

Mention some articles made of brass.

Write five lines about tin.

Write five or more lines about coal.

Write what you know of iron, gold, silver, copper, lead.

Lesson XLII.

MORE ABOUT THINGS FOUND IN THE EARTH.

We have seen that there are many kinds of

A GRANITE QUARRY.

metals. There are also many kinds of stone. Those which are strong and do not crumble by exposure are useful for building. The place from which stones are taken for building is called a *quarry*.

The more common stones are granite, sandstone, limestone, marble, and slate.

We will first examine a piece of *granite*.

How hard and firm it is! What a beautiful clean surface when polished!

Granite is used for steps, for paving streets, and for sidewalk curbings. Are houses ever built of granite? Can you think of other uses of granite?

Why is granite used for these purposes? It is easily shaped. It is hard enough to give strength. It is enduring. What does "enduring" mean?

This is a piece of *sandstone,* made of little grains of sand. It will crumble more easily than granite. What does "crumble" mean? Brownstone, used in building, is a kind of sandstone.

And this is the common gray *limestone* of which lime used in building is made. The large oven in which lime is burned is called a lime-kiln. Did you ever see one? Can you tell how the lime is made?

Here are three pieces of *marble.* This piece is pure white. This is colored. It is marked by many strange forms, as you see in your mantel-pieces and table-tops. In this piece, you see many colored spots—mottled it may be called.

Marble is beautiful when polished.

In what different ways have you seen

A MARBLE QUARRY.

marble used? What parts of furniture are sometimes marble? Why is it suitable for this? Is marble ever used for building houses? Do you think it would be good for that purpose? Why? Which, do you think, is the best of all building stones? Why?

Marble and granite are the most beautiful and enduring of all building stones.

Chalk is a variety of limestone. Could it be used as a building stone? Is chalk harder or softer than other stone?

You need not to be told the name of this dark stone. You could not get along well in school without *slate*. Slate is easily split into thin plates, and has a smooth, firm surface.

Slate is used to write on. It is used in house building. What part of a house is sometimes slate? Think of other uses. Why is it useful for these purposes?

We must not forget *brick* in our talks about things that come out of the ground. Brick is not found in the earth, as the metals and stone are found; but it is made of clay, which is itself a part of the ground.

Have you ever seen a brick-yard? What are some of the uses of bricks? What is the man called who builds houses of bricks?

Is glass taken out of a mine or quarry? No; but glass is made from sand, which is also a part of the ground.

In laying brick or stone, the mason uses *mortar*. Mortar is made chiefly of lime. Lime is made of stone which comes out of the ground.

If possible, visit mines and quarries. Take careful notice of all you see, and on your return to school tell what you have learned.

Lesson XLIII.

HOW PEOPLE LIVE AND WHAT THEY ARE DOING

"DID YOU EVER HEAR OF PEOPLE WHO LIVE IN SNOW HOUSES?"

Can you think of anything used in building houses that does not come from the earth?

Do all people have large, fine houses of brick or stone to live in? What is a tent?

WIGWAMS.

A wigwam? Who live in huts? Did you ever hear of people who live in snow houses?

HOUSES BUILT OF BAMBOO.

In some places houses are built of bamboo. Bamboo is a kind of cane that grows in warm countries.

What building is now going up in this place? Tell the use of stone, brick, mortar, iron, tin, lead, and glass in building the house.

"WHO LIVE IN HUTS?"

Where and how are they obtained?

We could not live without food. We must also have clothes to wear and houses to live in. Besides these, we need schools, books, and churches, which make us wiser and better. Now, if you think a little, you can name many other things which we need to make our homes beautiful. To supply us with all of these things, men must do many different kinds of work.

Where does the food we eat come from?

9

We get most of it from plants. Wheat, corn, peas, and beans are seeds of plants. Almost all our bread is made from wheat. Beets, turnips, and radishes are roots of plants. Lettuce and cabbage are the leaves of plants.

"OUR BREAD IS MADE FROM WHEAT."

Apples, peaches, pears, and other fruits grow on plants. All these we use for food.

Plants also supply us with material for clothing. Some clothes are made from cotton;

cotton grows in the pod of a plant. Some clothes are made from linen; linen comes from flax, which is a plant. Hats are made from straw; straw is the stem or stalk of plants.

Now, these plants, which supply us with so much of our food and clothing, do not grow of themselves.

The ground must be plowed, the seeds planted and taken care of while growing. So, outside the city, you may see a great many people at work raising grain, vegetables, and other plants. This occupation we call *agriculture* or *farming ;* the people we call farmers.

Animals, as well as plants, furnish much of our food. All meat comes from animals. We get milk from cows. From milk we make butter and cheese.

Animals also supply us with clothing. Many articles of dress are made of wool. Wool, you know, grows on the sheep. Shoes and kid gloves are made of leather. Leather is made from the hides of cows, sheep, oxen, and goats.

But animals could not live and grow if people did not carefully raise them.   In the

"HERDS OF COWS AND OXEN FEEDING."

country, you may see flocks of sheep and herds of cows and oxen feeding on the fresh sweet grass of the pastures.   Those animals are called stock.   The business of those who raise them is called *stock-raising.*

Most farmers raise cows, horses, and other animals.   Which land does the farmer use for pasture? What is a pasture? What is a meadow?

Grazing means feeding on grass. What animals have you seen grazing? Does a dog graze? A cow?

Mountains, so rough and rocky, are not good for farms and gardens. But many of them contain coal, on which millions of people depend for heat and light. In mountains, too, we find iron, which is more useful to us than gold and silver.

"A MINE IS LIKE A GREAT CAVERN."

To **get these,** thousands of men are at work in places called mines. A mine is like a great cavern. There is neither sun nor sky. Torches and lamps give the only light the miners have to see by. The air is damp and close. I suppose you would not like to work in such a place. Yet great numbers of persons are employed in *mining.*

How is coal taken out of a mine? What are the dangers of coal-mining? Try to find answers to these questions for yourself. If necessary, your teacher will help you.

In some parts of the country are forests of pine, oak, and other trees. Some of these forests are so large we might travel for days or weeks through them. From trees we get lumber. Lumber is needed for building houses and ships, and for furniture. So a great many men are employed in cutting down trees and preparing the wood for use. This is called *lumbering.*

The lumbermen go into the woods in winter, and build themselves little huts to live in. All through the winter months they work in the woods from sunrise to sunset, felling the best trees and cutting them into

logs. Then they haul them over the snow-
covered ground to the frozen streams, and
pile them upon the banks.

"THEY WORK IN THE WOODS."

Here the logs must rest
till the snow and ice have
melted and the streams
are full. Then they are
floated down to the great saw-mills, and cut
up into boards, laths, shingles, and other
kinds of lumber.

What is a forest? Name some forest trees
that grow near your home.

The sea yields much that we eat. Some parts abound in codfish, mackerel, and herring. Sardines, the little fish that come in boxes, are also found in the sea. It is the business of thousands of people who live near the ocean to catch fish, salt them, and pack

"IT IS THE BUSINESS OF THOUSANDS OF PEOPLE."

them, to send to those who want them for food.

Have you ever seen the ocean, or eaten any of its fish?

Name some fishes found in fresh water.

Name some kinds of fishes found in waters near where you live. How may they be caught?

Lesson XLIV.

MORE ABOUT WHAT PEOPLE ARE DOING.

In the city or town we shall find many of the people busy about something else than the occupations we have learned. What do you suppose it is?

If you go about the city, you will see large buildings several stories high, with long rows of windows, and great smoking chimneys. These are mills or factories, full of machines in motion doing their work almost like human beings.

The people who work in them make almost everything that is needed for our use. Wheat is changed into flour; cotton, into thread, fine muslins, and pretty calicoes; leather, into boots and shoes; iron and steel, into plows, stoves, and cutlery; lumber, into wagons,

carriages, and all kinds of furniture. Other
articles which we must not forget are elegant
jewelry, all sorts of ornaments for parlors,
and beautiful toys which you admire so
much.

"BUSY MILLS AND FACTORIES."

It would take a long time to name a small
part of the things made in the busy mills and
factories; but think of the articles used in

your home, and you may be sure they are manufactured articles. You see, *manufacturing* gives work to many thousands of persons.

What is cutlery? Name some articles of cutlery.

We need many things which we do not produce. Other people need things which they do not produce. How can each obtain what he needs? By exchanging one thing for another. This exchange of goods, or buying and selling them for money, gives rise to another occupation called *trade*, or *commerce*. So many people spend their time buying and selling grain, vegetables, clothing, boots and shoes, or in sending them to places where they are needed.

On all the large rivers and lakes you may see boats going up and down, carrying goods from one part of the country to another.

Can you think how goods are carried from place to place where there are no rivers? In countries where few people live, goods are often carried in wagons and on the backs of animals.

I wonder how many people have to work to get food and clothing for us. Make a list of all the occupations you can think of. Perhaps you can think of other occupations we have not named. Is dressmaking an occupation? Teaching? Which occupation would you prefer? Why?

If you think, perhaps you can tell why men do different kinds of work. What people do to make a living, depends very much upon the place they live in. For men almost always do that kind of work that pays them best for their labor.

Those who live where the land is rich and level will raise grain to make flour, or cotton and flax to make clothing. Some people among the mountains work in the mines. Some keep cows for their milk and butter, and sheep for their wool; for the hills and many of the mountain sides afford excellent pasture. People who live near the sea will be apt to catch fish along the coast, or engage in trade upon the water.

Employments in the city differ widely from

those in the country. Here, as we have learned, most people make their living by working in factories, or as merchants in buying and selling goods which come from all parts of the world.

All people do not live in the same way. Some people have no churches, schools, books, or factories.

What do people who live in this way eat? What do they wear? How do they spend their time?

Lesson XLV.

A REVIEW LESSON.

What kind of work is done by the people among whom you live? Are they farmers? How does the farmer make his living? Where does he sell the things which he raises? Where does he buy his sugar and tea and other things which he needs?

Do you live in a city? What are the chief occupations of the people? Do they work in shops or mills or factories?

Name some mills or factories in or near your city. What articles are made there? What manufactured articles are in the school-room? At home? What do you call the men who make these articles?

What kinds of goods are sold in the stores? What is a grocery store? A dry-goods store? A shoe store? Where did the things in these stores come from? Which were made in your city? Which were brought from other places?

What railroads or canals are in the city? Do boats come to the wharves? What do the boats or railroads take away? What do they bring in return?

Write the following:

Farmers raise (write the names of all the things you can think of).

Miners dig _____ out of the earth.

Quarrymen dig _____ from the quarries.

A shoemaker makes _____.

A blacksmith makes _____.

Merchants buy and sell _____.

22746051R00087

Made in the USA
Middletown, DE
07 August 2015